Dieter Mende
EEZ Energy, Energy industry, Future energies

The energy turnaround:
Perhaps seeming paradoxical at first?

Current positions.

The comprehensive bracket with environmental protection against climate change.

The globally booming image.

If someone tells you that they can explain the energy transition to you within a few minutes, then you should be extremely skeptical.

Thinking EU/nationally and acting locally is not a contradiction, but rather a dynamic energy policy.

Dieter Mende

Dieter Mende
EEZ Energy, Energy industry, Future energies

The energy turnaround:
Perhaps seeming paradoxical at first?

Current positions.

The comprehensive bracket with environmental protection against climate change.

The globally booming image.

Impressum

© 2024 **Dieter Mende**, EEZ Energie Energiewirtschaft Zukunftsenergien
www.eez-mende.de

Further contributors:

Antje Mende,   LIKES Layout – Impuls – Konzept – Entwurf – Style;
Moderne Medien-, Text- und Bildberatung

Herstellung und Verlag: BoD – Books on Demand, Norderstedt

ISBN: 978-3-7583-3172-5

Inhaltsverzeichnis

Prologue

My motivation for creating reports and books is, on the one hand, a passion for highlighting the opportunities and possibilities in the potential grid of the energy transition with hydrogen as an energy carrier and, on the other hand, the ambition to build and expand a hydrogen infrastructure with the advertising of cross-industry service providers, with the identification of sustainable contributions and with the resulting and complementary skills.

The target group of this book is broadly diversified; the book is aimed at readers who are less interested in technology, as well as students and teachers, politicians, entrepreneurs and technicians.
In order for this to succeed, the basis is created at the beginning of the book so that everyone can recognize the initial situation of the energy transition.
The course of the book is therefore designed in such a way that it does not demand too much basic technical understanding from readers who are less interested in technology, so that the course of the book does not appear boring for entrepreneurs and technicians.
As a result, the first chapter should provide readers with a common basic understanding of the energy transition. The second chapter gives readers a holistic insight from the sources, the regenerative paths of energy generation, to the sinks, the mobile and stationary applications.
Lassen sie sich davon begeistern, dass Veränderungen sehr viele Chancen ermöglichen, dass Veränderungen ohne Übertreibung sehr spannend sind und dass Veränderungen viel Begeisterung für die Zukunft auslösen können.

Please do not allow yourself to be unsettled by the deliberately created confusion on the part of the lobby against an energy transition, as the lobby against an energy transition deliberately presents incomplete arguments.

The energy transition must be viewed holistically so that the interrelationships that have a significant influence on climate change and environmental impacts can be recognized.
The energy transition must be viewed holistically so that the sound technological expertise and infrastructure know-how can be optimally integrated into the existing energy markets and can also expand the existing energy markets.

In the course of the book, you will also learn about what are certainly the most exciting developments in the modern world with the challenges of today; on the one hand with a view to maintaining the security of energy supply for people, and on the other hand with a view to the many opportunities for future generations.
The energy transition is much more than just increasing the use of renewable energies!
The energy transition is a job engine.

If the energy transition is to succeed and if the agreed climate targets are to be met, there is no alternative to the immediate expansion of renewable energies.

Energy transition: The current position

Since the publication of the EU climate reports, there has been international consensus that the energy transition is important. However, the paths of the energy transition are still a trigger for sometimes heated discussions today.

Even with the expansion of the grids for electricity in the direction of "copper plate Germany", full electrification has an implementation problem: the important energy storage systems are missing.
The stability of the electricity grids depends on the balance between the supply of electrical energy and the demand for electrical energy.

In addition to an increasing energy mix, the energy transition is also the transition from bound energy, bound in coal, oil and natural gas, to an increasingly regeneratively generated electrical energy.
With the planned wind turbines alone, we are already in the gigawatt range of electrical energy.
Batteries alone are no longer an option for storage on this scale.

The advocates of full electrification have not only moved out of the hedgehog position in view of the very high costs of grid expansion and are more willing to talk, because in many places a holistic view of the energy transition in connection with climate change and environmental protection has set in.

The energy transition must be viewed holistically, from the source (the regenerative paths of energy generation) to the sinks (the mobile and stationary applications). With Power-to-X, the energy transition links the previously separately considered paths of electricity with the paths of gases, as well as with the paths of heat generation and with the paths of fuel production. The energy carrier hydrogen is the comprehensive bracket for the success of sector coupling.

However, the energy transition is also producing numerous intelligent solutions through digitalization.
Not only do smart grids support the opportunities of sector coupling with hydrogen, smart applications can also lead to a paradigm shift with a view to the applications.

Currently, in grid centrality, power plants follow the energy demand in the electrical grids with energy generation. With digitalization, however, applications can also follow a spontaneous, higher supply of regeneratively generated energy in the grids by programming those systems that do not have to run continuously.

Two classic examples in the private sector could be:
+ e.g. the washing machine, which is programmed before leaving the house and starts primarily when the radio impulse comes from the network operator, which starts secondarily when a programmed target time is to be met.
+ e.g. the storage heater, which charges primarily with electricity when a particularly large amount of "green electricity" is being fed in by the wind and/or the sun, which charges with electricity at other times.

For a very long time, the energy infrastructure in Germany was characterized by grid centrality with a focus on the generation of electricity in power plants. The increasing decentralized and renewable energy generation resulting from the energy transition does not have to contradict this existing energy infrastructure; on the contrary, the hydrogen user center h2herten shows that the energy transition is a supplement to the current energy infrastructure.

With the energy transition, virtual power plants are being created by coupling renewable energy generation through digitalization.
The h2herten hydrogen user center can be operated autonomously as a stand-alone solution, but the center can also be operated in parallel with the grid and thus actively participate in grid operation.

The energy transition creates the potential for many self-sufficient systems to be combined in the electrical energy grid to form a mesh, similar to a fishing net, whereby these meshes in turn complement each other in the electrical energy grid. With grid management, intelligent, virtual power plants can be created that manage the stabilization of the grids and also ensure the security of energy supply.

The course of the book shows numerous opportunities arising from the energy transition, and the course of the book also confirms the energy transition as an internationally booming job engine.

The course of the book names persistent misinformation in the subject of the energy transition and shows you with the

top structured course how you can actually evaluate the energy transition
even as a layperson, how you can expose the deliberately incomplete or even deliberately false statements in the subject of the energy transition.

We have often heard the statement that we first have to find out whether electromobility with batteries or electromobility with fuel cells will prevail. It is true that electric mobility with batteries and electric mobility with fuel cells complement each other, just as combustion mobility with petrol and combustion mobility with diesel currently complement each other.

Nobody would think of questioning the simultaneity of diesel and petrol as a fuel for combustion mobility. So why the unnecessary discussion about the simultaneity of battery and fuel cell vehicles?

Because the lobbyists of the coal industry and the mineral oil industry are very well paid for their lobbying work with the aim of slowing down competing energy products in the energy markets for as long as possible.

No less often heard were the statements that security of supply with electrical energy can only be possible with grid centrality by maintaining energy generation by the power plants.

It is true that renewably generated energy with hydrogen not only enables decentralized stand-alone solutions, but also enables grid-parallel systems that can be coupled with intelligent grid management, including virtual power plants.

Just as often heard were the statements that Germany is an energy importing country and that energy in the form of hydrogen will also have to be sourced from abroad in the future because Germany cannot provide enough renewable energy of its own. It is true that hydrogen from abroad with its transportation infrastructure is considerably more expensive; every kilowatt hour of electricity generated in Germany helps to reduce energy costs. With the expansion of renewable energy generation through modern wind turbines in combination with the generation of electrical energy through modern solar systems, a considerable proportion of hydrogen production will be able to take place in Germany.

At the same time, you will experience what are probably the most exciting developments in the modern world with the challenges of today; on the one hand with a view to maintaining energy supply security for people, and on the other hand with a view to the numerous opportunities for future generations.

The book also lays the foundations for recognizing the opportunities presented by the energy transition.

If the energy transition is to succeed, there is no alternative to the expansion of renewable energies.

If the targets set with regard to phasing out coal are to be achieved, politicians must ensure that conflicting political decisions are avoided in the future.

On the one hand, the former German Federal Minister of Economics Peter Altmaier (CDU) named hydrogen as an important energy source for the success of the energy transition, but on the other hand, he slowed down the equally important expansion of renewable energy generation with stricter requirements, so that in many places where wind turbines are already in place, no more are currently allowed to be built.

These currently prevailing inconsistencies are the subject of admonishing people in Germany and also had a significant influence on the outcome of the 2021 federal elections.

One of the "three German words of the year 2019" is Klimajugend; in english: Climate Youth.

Picture: Dieter Mende
Wind, solar, GEO, bio, hydro ... energy generation is increasingly regenerative and requires suitable energy storage systems for an uninterrupted energy supply.

Energy transition: perhaps seemingly paradoxical at first:

Decentralized power supply is indeed also a paradigm shift, triggered by the energy transition.
However, the energy transition is not in contradiction to the current energy infrastructure; on the contrary, the energy transition combines the current energy infrastructure with numerous additional possibilities.

The central generation of large quantities of electricity, which has been common up to now, takes place centrally, at one location in the power plant. When the electricity is fed into the grid, the centralized power supply takes place. Increasingly, electricity is being generated from renewable sources at distributed locations, decentralized, using wind, sun, geothermal energy, hydropower and biomass; this can result in a decentralized power supply.

However, this is not necessarily the case. The regeneratively generated energy can be combined to form virtual power plants; the term virtual power plant may still seem abstract to you at this point, but the term virtual power plant will already become concrete to you in the first two chapters of the book.
Virtual power plants are implemented, for example, by linking combined heat and power plants (CHP) or residential and commercial district solutions with their very different energy requirements in intelligent grids.

For example, the hydrogen user center h2herten: the system management can actually be self-sufficient as an "island", but the system management can also be connected to the grid in parallel and thus make a valuable contribution to grid stability.

Grid stability enables an uninterrupted energy supply with a very high level of security of supply, while grid stability also has a stabilizing effect on energy prices thanks to the complementary district solutions in a virtual power plant.
The security of energy supply and the stabilizing effect on energy prices are the important basis of our economy and our standard of living.

The energy 3 leap is a term that is appearing more and more frequently in the energy transition dialog with the aim of reducing CO_2 emissions and is the conceptual bracket for the three goals:
+ 	increase energy efficiency,
+ 	reducing energy demand,
+ 	the expansion of renewable energy generation.

Fracking (hydraulic fracturing) has often emerged as a potential opportunity to extend the reach of fossil fuels, especially around the millennium. Fracking is a technical process by which the rock deep below the earth's surface is broken up by pressure generated by injecting a chemical fluid.

As the chemical fluid is injected into the highly compacted rock deep below the earth's surface, cracks are created; the existing cracks are widened.

The purpose of fracturing the rock is to enable access to the oil and natural gas contained in the rock. The problem with fracking is, on the one hand, the risk of contaminating the soil and groundwater with oil, natural gas or the chemical fluid itself, which is used to build up the pressure.

Another problem with fracking is the risk that the pressure that triggers the fracturing of the rock can also have geological effects, including shock waves.

Fracking also poses the problem that toxic fluids used to break up the rock remain in the ground and pose incalculable risks, because it is well known that the earth does not have an unmoving mantle, but is subject to constant change over decades and centuries due to plate tectonics.

Fracking also poses the major problem that methane escapes and further exacerbates climate change.

In addition, the actual goal of fracking, the extraction of crude oil and natural gas, does not fit in with the goals of the energy transition:
+   reducing climate change,
+   environmentally friendly energy production,
+   affordable energy production,
+   global support for poorer countries.

The energy turnaround is also the path from materially bound energies, bound in coal, oil and natural gas, to regeneratively generated electrical energy paths.

At this point, we must first clear up the unfortunately persistent misunderstandings, including the idea that we only need to expand the electrical grids and the problem will be solved with full electrification.

a) The energy transition is not the "problem", the paths of the energy transition are the solution to a conflict with a view to environmental problems, which include climate change.

b) The renewable generation of electricity is variable, as the intensity of the sun is not always the same, as the wind does not always blow the same way; the term for this is volatile.

c) An unlimited amount of electrical energy cannot be fed into the grids; the result would be an overload with the collapse of the electrical grids.

At this point, I would like to recommend a short film that, with a running time of just 1.41 minutes, skillfully and amusingly illustrates the requirements for a sustainable energy infrastructure.
You can watch the short film on YouTube by entering the German words in the search box: Nachbarn unter Strom. As the short film does not use any language, it is internationally understandable.

With the current energy infrastructure, the energy is transformed to the different voltage levels in order to reduce transmission losses.

The HVDC high-voltage direct current transmission line is a term used in the discussion about the energy transition to describe a line that enables the transportation of electrical energy in the form of high-voltage direct current. Electrical energy can be transmitted over long distances in the form of high-voltage direct current with significantly less energy loss. However, even with this method of transmitting electrical energy, there is the problem of a lack of energy storage.

Picture:
Copper plate Germany vs. hydrogen. The energy transition is also the move away from material-based energies, bound up in coal, oil and natural gas, towards increasingly renewable electrical energies, e.g. solar and wind power. The photo was taken from the Hoheward spoil tip in Herten, Germany.
Photographed by Dieter Mende, EEZ Energie Energiewirtschaft Zukunftsenergien,
Picture consulting by Antje Mende, LIKES Layout Impuls Konzept Entwurf Style.

Full electrification: in view of the energy transition, the question has arisen as to how energy can be transported in a climate-neutral and environmentally friendly energy infrastructure.

In this context, it is considered that with regard to the fossil, material-based energy supply, bound in coal, oil and natural gas, the energy transition lacks energy storage.

Due to the lack of sufficient energy storage, consideration was given to what a regeneratively generated electrical energy supply could look like. As a result of the consideration by the representatives of the electricity producers, the massive expansion of power lines was proposed, which led to the idea that the energy supply in Germany would only be provided by electricity and no longer by gas.

Due to the lines for electrical energy that would then actually be necessary, it was noted that Germany would then very soon have to look like a copper plate.

Copper plate is a term that is cropping up more and more frequently with regard to the energy transition and has arisen from the question of how energy can be transported in a climate-neutral and environmentally friendly energy infrastructure. In this context, it is considered that, in view of the previous fossil and material-based energy supply, the energy transition lacks energy storage.

As a result of the consideration by the representatives of the electricity producers, the massive expansion of the power lines was proposed, which led to the figurative idea of a copper plate Germany.

Hydrogen as an energy carrier, energy storage is the favored alternative.

With regard to the energy transition, the dark doldrums are the simultaneity of the lack of energy generation from the sun, e.g. at night, and the lack of energy generation from the wind.
The dark doldrums are contrasted with the previously stored electrical energy with hydrogen; hydrogen that was previously produced with the regeneratively generated energy that the energy suppliers' grids could no longer absorb.
In principle, the energy transition means that electricity generated from renewable sources should always be used directly! Up to now, wind turbines have been curtailed in their generation capacity to prevent overload situations from arising in the grids. The potential is created with the previously unused electrical energy stored in hydrogen:
+ the wind turbines whose generation capacity has been reduced to date already exist and therefore do not generate any additional system costs.
+ Due to the curtailment of the generation capacity of the wind turbines, the electrical energy that could potentially be generated is irrevocably lost, which results in a significant reduction in added value. This situation triggered the introduction of the EEG surcharge against the background: why invest in a wind turbine if the electrical energy that can basically be generated cannot be fully produced and sold?

We therefore have the situation:

+   that with this hitherto irrevocably lost electricity, the curtailment of the wind turbines with the generation of hydrogen storage does not result in additional costs for the energy generation itself, nor does it result in additional costs due to the existing plant itself.

+   the success of the energy transition in the grids means that energy generated from renewable sources must have priority over energy generated in power plants. This often leads to situations, triggered by unexpectedly high energy generation from the sun and/or wind, in which too much electrical energy is fed into the grids because the power plants cannot be reduced in output so quickly. Until now, it has been possible to transfer the excess electrical energy to the neighboring countries; the neighboring countries pay very well for this decrease in electrical energy.

This can lead to a truly paradoxical overall situation:

+   that money has been paid, initially for the construction of a wind turbine,

+   that money has been paid for the generation of electrical energy with the wind turbine,

+   that grid problems arise and with the shutdown of the wind turbine, with reduced income,

+   that grid problems arise and therefore a lot of money has to be paid for passing on the electricity.

So why should the continuation of the energy transition still make sense in view of the list of points on pages 20 and 21?

+ Because it is once again clear that a sensibly implemented energy transition must be considered holistically from the source, the increasing renewable energy generation, to the sinks, the mobile and stationary applications.
+ Because hydrogen energy storage has not yet been integrated into the energy infrastructure and the integration of hydrogen energy storage will make the energy transition a success.
+ Because the technology know-how and the associated infrastructure know-how will become a global job engine.
+ Because a dark doldrums, the simultaneity of the lack of energy production with the sun, e.g. at night, and the lack of energy production with the wind, with the hydrogen energy storage system is not an argument contrary to the implementation of the energy transition.
+ Because the lobby, which is acting against the energy transition with deliberately incomplete and/or false statements, must recognize with a holistic view that the energy transition with the energy storage hydrogen actually has a cost-reducing effect in addition to the effects of environmental protection and in addition to the meaningful contributions to avoiding the progression of climate change.
+ Because bringing reality closer to the claim actually triggers joy for the future.

Yes, there have also been a few learning processes with the start of the energy transition; however, there has also had to be a learning process with every new technical idea in the past. Or does anyone really believe that the light bulb developed by Thomas Alva Edison was the result of the very first approach? The steam engine was given a decisive improvement by James Watt in 1769, whereupon he was granted a patent.

Please try to see a "mistake" as a decision-making aid. In the most positive sense, "mistakes" are first and foremost results; results are the basis of all developments.
I interpret as a real mistake what would be prosecuted by the public prosecutor's office under the Criminal Code. I interpret antisocial behavior such as egoism as a mistake. I interpret ignoring and/or denying climate change as a serious mistake.

Learning has taken place, for example, with regard to wind turbines, the "disco effect" is exemplary.
The "disco effect" is a light effect that could emanate from the first types of wind turbines at a few very unfavorable moments. The rotating rotor blades of the first types of wind turbines did not yet have a matt coating, so that sunlight could be reflected from the rotor blades if the sun was at a certain angle to the rotor blades of the wind turbine and if the wind turbine was turned in a certain direction by the wind at the same time.
With the rotation of the rotor blades, the reflections of the sunlight on one of the rotor blades only occurred for a brief moment, but with the rotation of the rotor blades, the reflections of the sunlight were recurring on the following rotor blades.

With increasing wind speed and the resulting increasing rotation of the rotor blades, the appearance of the reflections of the sunlight also increased at ever shorter intervals, so that this effect was given the name disco effect. The coating of modern wind turbines means that the disco effect is no longer possible.

Unfortunately, the rebound effect often appears in connection with undesirable developments in the energy transition with a view to saving energy. The rebound effect is an effect that reduces the intended goal through knock-on effects.

With a view to the energy transition, more energy-efficient lamps have been brought onto the market with the aim of reducing energy consumption in the lighting sector, both in the private residential sector and in the commercial and industrial sectors.

This has led to a change in people's behavior to the effect that the lighting is not switched off as soon as they leave an area of the home or office, for example, because the more energy-efficient lamps supposedly result in much lower energy costs.

As a result, the significant increase in lighting duration and people's misbehavior has noticeably reduced the goal of reducing energy consumption.

Sufficiency is also a concept that is increasingly appearing in the energy transition dialog, calling for a way of life that is sustainable without reducing the quality of life.

The call for a sustainable lifestyle includes, for example, switching off energy-consuming devices such as PCs when you are not at home and switching off energy-consuming light sources when you actually leave the area. Sufficiency is not a term that is intended to have a prescriptive effect, but rather to lead to a prudent way of life.

Exergy is the term for the overall energy process of a system that can perform work; an example of poor exergy are the condensing boilers in residential buildings.
Although condensing boilers are very good in terms of energy, as they use the condensation heat in the exhaust gas, they are miserable in terms of exergy, because the combustion of heating oil or natural gas generates a firing temperature of 1,000°C in order to maintain a flow temperature of just 70°C for the radiators. 70°C flow temperature of 1,000°C firing temperature is not energy efficient.

Energy is the result of the sum of exergy plus anergy, whereby anergy is the term for the part of the energy that cannot play a role in an energy process.
With regard to the energy transition, special attention must be paid to exergy; there is a lot of potential for optimization here.

With the increasing renewable energy generation in the energy transition, the term grid parity appears; this with a view to costs.

Grid parity is achieved when the cost of generating renewable electrical energy, e.g. from wind or solar power, is the same as the cost of conventional electrical energy generation, e.g. from coal-fired power plants.
Grid parity is achieved when, for example, the cost of generating your own electricity with solar panels on your roof is the same as buying the same amount of electricity from the grid.

With the increasing renewable energy generation in the energy transition, the term seems to be inherent in the system.
The lobby against the energy transition has often cited the technically modern technology that makes conventional energy generation more effective, with which the efficiency of energy generation is higher, in order to avoid the expansion of renewable energy generation. The initial spontaneous reaction with the question of why this technically modern energy generation is not yet being implemented in reality is quickly followed by the realization that conventional energy generation paths are already exploiting the maximum possible technical potential. This is justified by the technical term "system-inherent".
System-immanent means that all technical goals are limited in their realization by the scientific laws of physics and chemistry; only within these limits can the desired technical efforts for optimization be implemented.

The term system-inherent is usually used when the overall objectives are very complex because the implementation of the energy transition is interdisciplinary.

The course of the book demonstrates in numerous ways that the energy transition must be viewed holistically:

+	From energy generation to applications.
+	Coupling the previously separate paths of electricity, gas and heat with sector coupling: Power-to-X.
+	Using smart grids to combine decentrally generated energy into grid meshes and combine these into a virtual power plant.

It is not uncommon for the loudest critics with a negative stance towards the energy transition to become rather nervous on closer inspection, because the critics often engage in paid lobbying with the aim of slowing down the envisaged energy transition. In July 1995, I initially started regionally with a focus on hydrogen energy storage; for the few polite people I had the opportunity to speak to, I was the visionary, for the others I was more of a crank. However, as early as 2002, the sustainable potential in the Emscher-Lippe region made it possible to convince people nationwide. Today we know that those who originally intervened against the start of the energy transition with hydrogen energy storage, with deliberately incomplete and/or deliberately false statements, are themselves striving to enter the hydrogen market with their own competitive ideas.

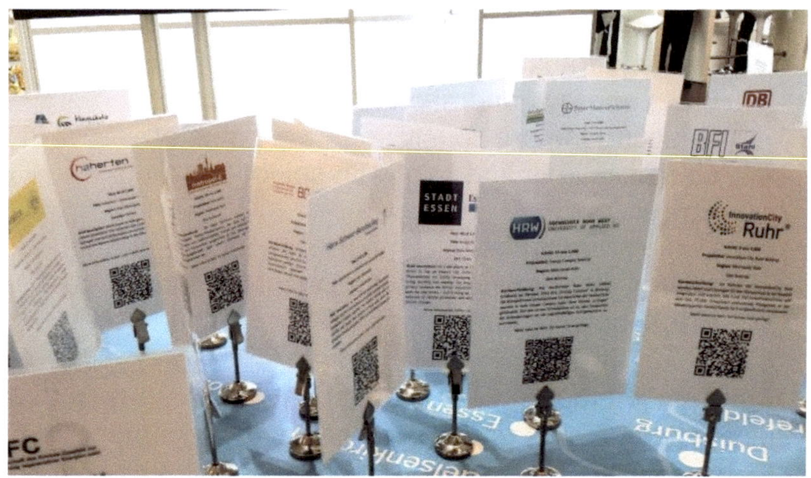

Picture: Dieter Mende
KlimaExpo.NRW exhibition during Energy & Water 2015 in Essen (Germany);
EEZ Energie Energiewirtschaft Zukunftsenergien

Picture: Dieter Mende
Electric drives with batteries and fuel cells complement each other, just as gasoline and diesel combustion engines currently complement each other. Hydrogen filling station from Air Liquide and charging stations from Tesla in Kamen (Germany);
EEZ Energie Energiewirtschaft Zukunftsenergien

Energy transition: explaining the tangents:

The Intergovernmental Panel on Climate Change is an institution set up by the United Nations; the IPCC is based in Geneva. The IPCC publishes regularly updated assessment reports and reports on the state of knowledge regarding climate change in the IPCC Assessment Report. The governments of the United Nations can use the information from the IPCC to make their own recommendations for action; the IPCC itself does not carry out research, but rather deals with the completed studies and publications and provides a scientific assessment of the results.

With regard to the energy transition, environmental responsibility is the driving force behind efforts to expand renewable energy generation; an ideology is the resulting goal.
With regard to the energy transition, an ideology arises from a world view of what energy generation should ideally look like in the future:
+   climate-neutral,
+   environmentally friendly and
+   affordable for all people.

In fact, the energy transition is an environmental necessity. The sustainable ways to achieve the energy transition are to be found in reality between ideology (the theoretically most valuable approach) and openness to technology.

Technological openness is an often recurring term in the energy transition dialog, which demands that there should be no discrimination against the general technical possibilities for generating energy.

As a result, the path to future technologies runs via bridging technologies; in this context, Power-to-X is a sector-coupling technology option that couples the electricity, gas and heat sectors.

The demand for technological openness may well be at odds with ideology. However, with regard to the energy transition, there are in fact also concerns that restrict technological openness.

Above all, the internationally declared climate targets restrict the openness of technology. If a technology cannot contribute to the internationally declared climate targets, this can become an exclusion criterion.

If a technology even works against the internationally declared climate targets, this creates a justified exclusion criterion; the decision to phase out coal is such an environmental policy outcome.

The concept of decarbonization gained global political significance with the G7 summit in 2015.

The world's seven largest economies set the goal of achieving complete carbon neutrality for the global economy by 2100. With this ambitious goal of complete CO_2 neutrality by 2100, the timeframe may still seem long, but in terms of global implementation, it is very ambitious.

Sustainable energy generation: in addition to the terms environmentally friendly and climate-neutral, the term sustainable energy generation keeps cropping up in the energy transition dialog. This refers to the promotion of renewable energy generation, e.g. with wind and solar power.

Burning fossil fuels such as coal, crude oil and natural gas not only releases carbon dioxide (CO_2), which is harmful to the climate, but the energy source itself is lost through the combustion process:

+ With the electric current, the technical plant electrolysis produces the hydrogen and oxygen from the water.
+ Conversely, the fuel cell plant uses the hydrogen and oxygen to generate electricity and water.
+ The electrolysis plant works with electricity that has previously been generated from renewable sources, e.g. the sun or wind.
+ These processes can be repeated endlessly, making them sustainable, whereas coal, natural gas and crude oil are lost in the combustion process.
+ Sustainable energy generation also includes climate-neutral and environmentally friendly energy storage, which means that hydrogen energy storage completes the energy transition.

Sustainable energy generation with the points shown above is combined in an EKS energy complementary system.

In the energy transition dialogue, an EKS energy complementary system is the term for a sustainable energy infrastructure system, as demonstrated in the h2herten hydrogen user center. The EKS is the answer to the question of decarbonization, the answer to the search for a sustainable energy infrastructure without burning fossil fuels.

Picture:
Dieter Mende

In the foreground:
The entrance to
the hydrogen user
center h2herten.

Behind it:
the existing
buildings
of the former
colliery
on Ewald.

In the
background:
the silhouette
of the Scholven
power station
in Gelsenkirchen
(Germany).

EEZ Energie
Energiewirtschaft
Zukunftsenergien

The Stone Age did not end because there were no more stones and the energy transition will not come when there is no more oil/natural gas.

In the course of discussions about the energy transition, the term depletion mid point (dmp) has repeatedly cropped up due to the question of how urgent the implementation of the energy transition really is in terms of the range and availability of fossil fuels such as coal, oil and natural gas. depletion mid point (dmp) is the term used to describe the point in time at which half of the original energy reserves, or half of the total potential, has been extracted. The assumptions as to when this point has been reached differ; differentiated on the one hand by the consideration of basic availability, such as tar sands, and on the other hand by the consideration of the economic viability of their extraction.

At the end of this consideration are the energy costs, e.g. for gasoline at the filling station, e.g. for heating oil and natural gas for heat generation.
In principle, the recycling of tar sands is technically feasible, but the cost of petrol, for example, would be so high that petrol would become a luxury item and would no longer be available to the general public for everyday car use.
Against this background, there is a growing number of assumptions that humanity has not only already reached the dmp depletion mid-point with regard to fossil fuels, including the current technical possibilities for extraction, but has in fact already passed it.

If we now also consider the increasing energy requirements of emerging and transition countries, the need for an immediate start to the expansion of renewable energy generation becomes abundantly clear.

Bridging technologies and future technologies complement each other with Power-to-X. Bridging technologies, this term is used when existing technologies are integrated into renewal processes with the aim of developing future technologies.

With regard to the energy transition, this could be power-to-gas. The goal of moving away from material-based fossil fuels such as coal, oil and natural gas towards renewably generated energy is achieved with hydrogen energy storage, a sustainable and environmentally friendly technology that has a climate-neutral impact.

As long as there are not enough hydrogen storage facilities available to enable hydrogen production to start on a large scale, feeding hydrogen into the existing natural gas grid is a technical possibility; natural gas technology will become a bridging technology.

In electromobility, hybrid technology is also a bridging technology until sufficient fuel cell vehicles and hydrogen filling stations are available. In terms of technical implementation with a view to the range of the vehicles, driving with electricity from the battery and driving with a petrol engine complement each other.

Bridge technologies are gradually followed by exnovation. Exnovation is the term for the opposite development to innovation.

With regard to the energy transition, exnovation means that existing technologies that cannot meet the agreed international climate targets are being systematically dismantled because these technologies are not climate-neutral and/or not environmentally friendly; the coal phase-out is an example of exnovation.

The German Federal Immission Control Act (BlmSchG) was passed with the aim of preventing immission values from being exceeded,

+	which may affect humans, fauna and flora,
+	which may affect the environment and the soil,
+	which can affect the ambient air and water.

The Federal Network Agency (BNetzA) is a German federal authority within the remit of the Federal Ministry of Economics. The Federal Network Agency is responsible for the electricity network, the natural gas network, the telecommunications network, the postal service and the railroads and is based in Bonn. The Federal Network Agency ensures compliance with the guidelines with a view to fair competition in the networks. In Germany, energy policy is an integral part of economic policy and serves general economic and socio-political objectives. With a view to preserving the natural basis of life and also with a view to the energy transition, all measures that also enable a reduction in emissions with the processes of energy generation and with the processes of energy applications play an important role in energy policy.

The greenhouse effect has long been an issue in Europe, even if it has not always been focused solely on energy production.
In the atmosphere, trace gases with an effect on the climate allow the short-wave solar radiation to pass through to the earth almost unhindered, but retain a large proportion of the long-wave heat reflection on the earth's surface.
In the Earth's natural balance between the short-wave solar radiation to the Earth and the long-wave heat re-radiation from the Earth, the natural greenhouse effect causes an average temperature on the Earth of plus 15°C; without the natural greenhouse effect, the average temperature on the Earth would be well into the deep freeze range, so that life as we know it on Earth would not be possible without this natural greenhouse effect.

With regard to the energy transition, there are man-made changes in the atmosphere which mean that the long-wave heat reflection from the earth no longer takes place as it used to. As a result, this development leads to global warming. The increasing emission of climate-impacting trace gases is primarily due to the combustion of fossil fuels.

Emissions are pollutants (gaseous, liquid or solid) that are released into the environment by a plant, a building, a factory or a means of transportation. Emissions are also defined as the release of heat, noise or radiation into the environment. The impact of emissions is referred to as immission.
Here there are limit values (maximum values) for the legally permissible release of pollutants into the environment.

Immission refers to the impact of emissions on soil, water, humans, fauna, flora or material assets. One measure of emissions is, for example, the concentration of a pollutant per square meter of surface area or per cubic meter of air.

Resilience is the ability of ecosystems to adapt to new conditions in order to preserve them and ensure their continued existence. With regard to the energy transition, it is clear that the ability to adapt to new conditions has only been successful up to a certain point, and that ecosystems are already clearly exceeding their adaptability with climate change.

Reaching the tipping points must be avoided to prevent irreversible global consequences with catastrophic consequences for life on earth.

Tipping points is a term that refers to the Earth's climate system with regard to the energy transition. Whenever the scientific fields of biology, physics and/or chemistry are examined more closely, the complex interrelationships become apparent. It is no different with the Earth's climate system.

Tipping points is a figurative term based on the sequence of falling dominoes. If the interrelationships of the earth's climate are seriously altered by human influence, so that a natural sequence of events changes, this irreversibly triggers subsequent reactions.

Researchers' publications often cite the melting of the perpetual ice or the collapse of the influence of the rainforests on the earth, triggered by progressive deforestation, as examples of tipping points.

The actual tipping points are numerous and can only be recognized with an intensive examination of the earth's climate. The increase in the earth's temperature is already showing the subsequent reactions with the regional drought months, with the regional heavy rainfall, with the significant increase in storm events.
We are observing these developments not only with increasing intensity, but also with increasing frequency.

The lobby against the energy transition often emphasizes that only the maximum value creation of existing industries can bear the costs of the energy transition. The fact that this is not the case was mentioned earlier in the book.

The lobby acting against the energy transition focuses on the fact that the earth's energy reserves should be used up first in order to then gradually move on to the development of the energy transition.

It is just that climate change is not aligned with the economic interests of the industries that make a lot of money from conventional energy production paths.

The energy reserve is the term used to describe the reserves of fossil fuels stored in the earth:

+ which are proven,
+ which are safely available and
+ which can be extracted economically with the current state of technology.
+ Which ones if world energy demand remains constant,
+ which with assumed constant world energy use,
+ which assume a linear increase in consumption with the currently known world energy reserves of crude oil and natural gas.

The range of the known world energy reserves of crude oil and natural gas is often assumed to be 40 to 60 years. Unfortunately, the assumptions of a linear increase in global consumption are not correct.

The lobby acting against the energy transition cannot stand up to its own stated positions in a serious analysis.

In addition to energy reserves, energy resources are proven stocks of energy sources, supplemented by assumed stocks of energy sources.

Energy sources that are currently not recoverable for technical or economic reasons are also included; energy resources are therefore not a basis for reliable figures.

The course of the book shows that:
+   the rising population figures,
+   with dwindling energy reserves,
+   are already triggering climate change,
+   so that the expansion of regenerative energy
        production must start for this reason alone.

In the field of renewable energy generation, there are two starting points for the owners and operators of wind turbines with the expiry of state subsidies:
+   Replacing the existing wind turbine with a new wind turbine with the aim that the new wind turbine also receives a subsidy from the state as compensation for the electrical energy that the wind turbine can in principle generate, but which cannot be absorbed by the grid of the energy supply companies: repowering,
+   or the expansion of the existing wind energy plant with an energy storage system, which stores the electrical energy that the wind energy plant can generate in principle but which cannot be absorbed by the grid of the energy supply companies: hydrogen offers enormous potential with Power-to-X.

By integrating the technical possibilities into existing systems, the innovation triggers the desired goal of better energy efficiency.
In mobility, for example, this includes recuperation as well as hybrid technology. Recuperation is the recovery of energy in an electric vehicle via the braking process. The conversion of kinetic energy into electrical energy during the braking process is followed by the storage of the generated electrical energy in the vehicle's battery.

The electricity feed-in tariff guarantees private operators of renewable energy systems, e.g. wind or solar, a legally regulated remuneration for feeding the electricity they generate into the public grid. With a view to the energy transition: The EEG Renewable Energy Sources Act was passed to ensure that operators of renewable energy generation plants are compensated for the financial loss if the electricity grids are unable to fully absorb the electricity generated.

The EEG Renewable Energy Sources Act came into force on 01.04.2000 as a replacement for the law on electricity feed-in. The EEG Renewable Energy Sources Act has also become important in view of the ongoing "chicken and egg debate".

To explain the term "hen-egg debate": there is the question of what came first, the hen or the egg?

Transferring this "chicken and egg discussion" to the various areas of the energy transition:

+ why should investors invest their money in a wind turbine if the wind turbine cannot feed all of its electricity that can in principle be generated into the grid of the energy supply companies and thus the maximum possible added value of the wind turbine is therefore not possible?
Conversely, why should energy supply companies expand their grids with energy storage systems for large quantities of renewable electricity if there are not enough renewable plants?

+ Also interesting: why should the automotive industry bring battery or fuel cell vehicles onto the market if there are not enough electric charging stations or hydrogen filling stations?

Conversely, why should electric charging stations or hydrogen filling stations be built if there are not yet enough battery or fuel cell vehicles to make their development and expansion worthwhile?

Ideology alone cannot move investors in the direction of the energy transition. Investors are primarily interested in the maximum possible added value of the current business areas.

In the past, the "chicken and egg discussion" has made it easy for the lobby against the energy transition to slow it down.

+ The Renewable Energy Sources Act (EEG) has had the successful effect of greatly increasing the construction and expansion of renewable energy generation plants.

+ Hydrogen energy storage has the successful effect that large quantities of electricity generated from renewable sources can be stored for times when not enough electricity can be generated by the wind and/or the sun, such as during the dark doldrums.

In Germany, the Renewable Energies Heat Act (EEWärmeG), with regard to new buildings for which the building application was submitted after January 1, 2009, stipulates that the guidelines for preventing heat loss must be complied with.

The German Renewable Energies Heat Act (EEWärmeG) also stipulates, with regard to fundamentally renovated public buildings, that a proportionate generation of renewable energy is guaranteed for the heating requirements of the building; the cooling requirements of a building are also covered.

Only when the energy transition is viewed holistically can the meaning of the laws be recognized.
Residual load is also a term in the energy transition that requires a great deal of attention.

+ On the one hand, the residual load quantifies the difference with regard to the actual electrical power fed into the grids of the energy supply companies from renewable energy generation, which is naturally volatile and fluctuating,
+ On the other hand, the residual load quantifies the actual power demand for energy in the electricity grids.
+ This difference is balanced out by energy generation in the power plants.

With the implementation of the energy transition, the aim is to gradually
+ replace nuclear and coal-fired power plants by expanding the generation of renewable energies
+ and the simultaneous expansion of storage facilities for electrical energy, e.g. with the production of hydrogen with the excess capacities; renewable electricity volumes that the grid of the energy supply companies could not absorb.

Overload is a spectre that the lobby against the energy transition has liked to paint in the past. With hydrogen energy storage, the lobby against the energy transition has been deprived of its basis.

Yes, as previously shown in the course of the book, grid stability does indeed depend on the balance between the feed-in of electrical energy into the grid and the take-off from the grid.

However, if the energy transition means that the quantities of energy that exceed the capacity of the electrical grid are stored, then the electrical grid of the future will be even more stable thanks to the virtual power plants mentioned earlier in the book with their mutually stabilizing meshes in the grids.

<u>So that you don't have to go back in the course of the book, as the term virtual power plant doesn't really catch on immediately for many readers, here is an additional explanation of the term:</u>
As the wind does not always blow with the same strength and the sun does not always shine with the same strength, energy storage systems are becoming increasingly important with the further expansion of renewable energy generation.

The excess of electrical energy is the overcapacity.

The virtual power plant combines the many individual, self-sufficient systems. These can be controlled from outside via a central control system.

This creates the possibility that, for example, a housing estate can be supplied completely self-sufficiently, creating a quasi island solution, but also the possibility that many island solutions can be operated in parallel with intelligent networking in the energy supplier's grid and complement each other with the possibility that the load in the grid may shift. All in all, the virtual power plants create an overall grid-stabilizing strategy, an intelligent grid management system.

Sustainable energy infrastructure is a term that comes up again and again in the energy transition dialogue and summarizes the points mentioned above.

If someone tells you that they can explain the energy transition to you in a few minutes, then you should be very skeptical!
The energy transition is not trivial!

The sustainable energy infrastructure of the future is the path away from fossil and material-based energy sources, bound in coal, oil and natural gas, towards a climate-neutral, environmentally friendly and also affordable energy source for large amounts of energy: hydrogen.

In this context, bridging technologies, Power-to-X, are a sector-coupling technology option that couples the electricity, gas and heat sectors.
The demand for technological openness is indeed a justified demand in the energy transition dialog.

However, technological openness can also contradict the energy transition. In the energy transition dialog, there are also issues that restrict technological openness; above all, the internationally agreed climate targets.

The two-degree target is an agreement reached in Paris as part of the UN Climate Change Conference.
The 194 member states as a whole also committed to this two-degree target for the first time in December 2010. The two-degree target aims to keep the increase in the global average temperature to well below 2°Celsius.

This closes the circle with the core statement:
The energy transition MUST be viewed holistically: from the source, renewable energy generation to the sinks, mobile and also stationary applications.

Picture:
In addition to Power-to-X, climate change also requires the coupling of Power-to-Agriculture.
Photographed by Dieter Mende, EEZ Energie Energiewirtschaft Zukunftsenergien,
Picture consulting by Antje Mende, LIKES Layout Impuls Konzept Entwurf Style.

Energy transition: the comprehensive bracket with environmental protection against climate change

The energy supply of the future is inseparable from climate change and environmental protection, but the belief in this has not yet arrived everywhere in society.

Challenging and reminding impulses:

What people can see with their own eyes appears real to them and they are not so quickly inclined to follow a purpose-driven argument, such as the arguments of the coal lobby. People are not so quick to succumb to the deceptive words of others when it comes to what they can see with their own eyes.

People are quick to doubt what they cannot see with their own eyes, such as the beginnings of climate change, which researchers and environmental experts warned about many years ago, and are also more likely to succumb to the deceptive words of others.

Does it always have to be the emerging crises and fatal consequences that trigger recognition? ... from which a rethink arises?

The explanations of the researchers and environmental experts have not provoked any contradiction, since they are visible:
The dust that the large erupting volcanoes emit into the environment has led to a drop in temperature in the affected countries in previous years due to the physical effects of the dust masses, with the result that agricultural harvests have been very poor.

The dust from the eruption of the Pinatubo volcano in 1991 is estimated at 10 cubic kilometers (km^3) of dust mass.
The dust mass actually cooled the atmosphere somewhat, but did not have the same drastic global impact as the volcanic eruption in Indonesia in 1816, which with its dust mass led to a year without a summer for many regions. The cooling of the atmosphere in 1816 had the consequence that the very poor agricultural harvests in many regions caused a threatening famine.

Nobody doubted the explanations of the researchers and environmental experts; the connections between the drastic global effects were both explainable and visible.

We can assume with a high degree of probability that the onset of climate change in 1991 counteracted the effects of the eruption of the Pinatubo volcano to some extent.

The past few years in Europe have been recorded as the warmest years worldwide since weather records began in 1781.

The global effects of global warming are evident. The low-lying tropical coastal regions are facing an additional threat.

The warmer oceans are pumping more water vapor into the circulating low-pressure areas, meaning that cyclones are occurring more frequently than in the past and are racing towards the coasts with noticeably greater force. In general, climatologists expect an increase in extreme weather conditions such as droughts, hurricanes and heavy rainfall in the future.

With the obvious consequences of climate change, many people have begun to accept the necessity of the energy transition.
Future-oriented companies are starting to secure competitive advantages in a sustainable energy infrastructure with the energy transition.

What can be the reason for politicians not taking a position in the interests of the population, but in the interests of economic interests?
Greed for even more money, which counteracts the ability to solve society's problems through thought and action?

Donald Trump, the former President of the United States of America, is a fitting example of this.

In 2019, the Swedish climate protection activist Greta Tintin Eleonora Ernman Thunberg was much ridiculed by then President Donald Trump for her efforts to draw attention to the climate problems in the USA.

In fact, however, Donald Trump's reaction made it clear that he was unable to provide a credible counter-position to the climate facts and that, in his helplessness, he took flight in a more than unattractive gesture of disdain.

The deniers of climate change cannot hold their own in serious, fact-based discussions because the results of studies worldwide are consistent and complementary.

When people lose confidence in politics, the loss of trust in the solution to society's problems is more than just an alarm signal.
Doubts about political measures can be dispelled through transparency.

However, if doubts arise with regard to the politicians themselves, the basis for solving problems is shaken, perhaps even severely damaged.

Against this backdrop, politicians need to pay more attention to their reactions to the questions and concerns of the population.

Armin Laschet, the former German Minister President of North Rhine-Westphalia and also a candidate for chancellor in the 2021 federal elections, has shown little sensitivity to the pressing questions of the concerned.
During the interview of Armin Laschet and Tesla founder Elon Musk with the press at the construction site of the new Tesla plant, both were confronted with the concerned question about the water shortage in Brandenburg.

It was not only the reaction of US entrepreneur Elon Musk that caused outrage among the concerned citizens' initiative; Armin Laschet, who was campaigning for the German chancellorship, did the same.
In a situation in which many people in Brandenburg are concerned about the water supply of the future, the interview would have been a good time for a dialog with transparency, with a view to the operational management planned at the Tesla plant, and also with a view to the water management of the future.

Instead, Elon Musk ironically referred to the rain that fell during the interview and responded in the manner of Donald Trump with a counter-question:
"... what's the water problem here?"
Elon Musk laughs and Armin Laschet laughs with him.

The important paths to sustainable infrastructure in the energy transition are not always immediately obvious.
That is why the questions of concerned people must also be given a lot of attention so that the path to the energy transition can become a common path with unanimous acceptance in society.

A future-proof infrastructure in the energy transition combines the energy issues of electrical energy, gas and heat, which were previously considered separately, with smart grids.

In recent years, the lobby of power plant operators has spoken with "siren-like chants" about the full electrification of the economy in deliberately incomplete presentations.

A holistic view of the energy potential clearly shows that coupling electricity with gas, heat and smart grids reduces the costs of the energy transition and at the same time generates great potential in terms of technical possibilities. Linked to this is the energy transition as a job engine.

Resisting the song of the sirens:

The name Odysseus is known from Greek mythology. Odysseus was given the rule of Ithaca by his father. In the course of one of his journeys, Odysseus passed the island of the Sirens and was aware of the danger that any man who succumbed to the lure of the Sirens' sweet song and headed for their island would be lost and die. Aware of this danger, Odysseus glued his traveling companions' ears with wax and had himself tied to the ship's mast, so that Odysseus and his traveling companions could safely complete their journey home to Ithaca.

With regard to the challenges of the energy transition, there are also tempting siren songs from the lobby, which often want to lure people into their market positions with deliberately incomplete contributions.

People should certainly not close their ears to the energy transition.

However, attention is required when promises are formulated with trivial-sounding answers to the many complex tangents of the energy transition.

A great deal of attention is required when quick successes are predicted for the many complex tangents of the energy transition.

The energy transition is not trivial!

The results of the studies on the issues of the energy transition from a holistic perspective actually show that the coupling of the electrical energy sector with gas and heat maximizes the technical potential.

In fact, the results of the studies on the issues of the energy transition from a holistic perspective also show that the energy industry is not being turned upside down at all, but rather that the very many expansion potentials optimize the desired result of the energy transition.

If, contrary to the results with regard to sector coupling, there is still no uniform understanding of the path to the energy transition, then the reasons for this can be found in the recognizably incomplete lobbying of the individual energy paths.

The idea of the consequences of full electrification has deliberately not been communicated, because if the energy industry wants to generate heat with electricity, the additional demand for electricity would be enormous.

Moreover, this is in a situation in which we have to realize that we are nowhere near being able to solve the challenges of phasing out nuclear and coal energy with the plants for generating renewable electricity that have been announced so far.

The siren call for full electrification lures us in with the simple and convenient-sounding claim that we only need to be able to provide enough renewable electricity and all energy issues will be solved, including mobility.

With the expansion of renewable electricity generation, CO2 emissions will fall and the goal of reducing the effects of climate change will be achieved. This sounds simple and coherent at first; however, the enormous costs for the power lines are concealed.

The resulting scenario has given rise to the term "Germany's copper plate"; a term that represents the foreseeable enormous costs.

The enormous amounts of energy in the gigawatt range that are to be generated by the planned wind turbines, which will then be stored using which technical solution, are being concealed.

In the gigawatt range, we really no longer need to talk about batteries.

The term "power-to-gas" was coined when considering the sector coupling of electricity, e.g. with natural gas; excess electrical capacity is used to produce hydrogen, which can be fed into the gas grid.

As the path of the energy transition is also the path away from material-based energy sources (energy bound in coal, oil and natural gas) towards regeneratively generated energies that are electricity-based, hydrogen is not only a classic energy storage medium, but also an energy carrier that enables sector coupling as a connecting element and triggers a variety of technical possibilities, which have been given the collective term Power-to-X.

Power-to-X includes, for example, the potentials:
+ power-to-gas,
+ power-to-heat,
+ power-to-liquids,
+ power-to-chemicals,
+ power-to-fuels.

Farmers are increasingly also becoming energy producers. With the increasing problems caused by climate change, Power-to-X also has the potential to
+ power-to-agriculture.

Globalization has expanded online trade.
With the increasing problems caused by climate change, Power-to-X also has the potential:
+ power-to-LML last mile logistics.

Climate change and the necessary drinking water.
Power-to-X includes with the increasing problems:
+ power-to-RDW recovery of drinking-water.

Der Blick auf die Zukunftstechnologien zeigt durch die ganzheitliche Betrachtung auch die Potenziale der Brückentechnologien.

Behind this idea is the possibility that, when the feasibility and implementation of a future technology in the energy transition is examined, a closer look at the future technology, which includes hydrogen, often reveals interesting and complementary potential. With the already existing components of the energy infrastructure, the paths into the energy industry with hydrogen can be flexibly designed on the one hand, and the paths into the energy industry with hydrogen can be realized directly on the other.

As a result, the bridging technologies lead to the energy transition using hydrogen as an energy carrier to anchor renewable energy generation in as many energy markets as possible right from the start.

Power-to-gas and methanation can be a bridging technology.

The question of how to store renewably produced hydrogen is a question often asked by those who want to slow down hydrogen in the energy transition, usually against the background of a lobbying mission.

Methanation is technically somewhat more demanding than power-to-gas, but methanation has long since ceased to be a technical novelty and is mature in its implementation.

With methanation, the green hydrogen H2 is not fed to a storage facility, but to the carbon dioxide CO2; the gases react with each other to form methane CH4 and oxygen O2. Methane is the main component in natural gas; methane can be used as an energy source and the end product of methanation is often referred to as artificial natural gas.

When looking at methanation, the question inevitably arises: Who comes to whom?
The hydrogen to the carbon dioxide, or vice versa?
In all probability, with a view to the environmental compatibility of carbon dioxide, the hydrogen goes to the carbon dioxide.

In view of the link between the energy transition, climate change and environmental protection, methanation can counteract the high costs that will be incurred if the carbon dioxide CO2 is to be disposed of; storage requires considerable technical effort and is also very expensive.

Carbon dioxide CO2 will continue to accompany us in many industrial processes in the manufacture of products in the future, e.g. in cement production.
If we want to continue to build with concrete in the future, which is to be assumed due to the lack of alternatives, we will also be dealing with the issue of CO2 in the future.
The production of cement is so CO2-intensive that cement production is said to account for up to 8% of annual global carbon dioxide emissions.

Whether the focus is on power-to-gas or methanation, the two processes have in common that they provide a central answer to the requirements that have arisen with the question of sufficiently large hydrogen storage facilities.

In the event that the energy industry were to immediately start producing large quantities of green hydrogen, assuming that there would initially be more hydrogen producers on the market than hydrogen users:

The gas network in Germany alone is over 500,000 km long and has an enormous storage potential with an annual transport of almost 1,000 billion kilowatt hours (kWh).

At 540 billion kilowatt hours, the electricity grid in Germany is actually a far cry from the gas transportation grid at almost 1,000 billion kilowatt hours.

In 2020, the German Federal Ministry for the Environment confirmed on its website with regard to Power-to-X (PtX) that green hydrogen and electricity-based fuels and raw materials can make an important contribution to climate protection.
The goal of decarbonizing the economy is of both national and international interest.

According to the Federal Ministry for the Environment, this also applies to the transport sector, where the direct use of electricity will probably not be possible in the future for technical reasons.
The storage of electricity for long-haul aviation and long-haul shipping cannot be achieved in batteries due to the large amount of space required and the very high weight of the batteries.

Here, too, it becomes clear that the formula for the success of the energy transition is:

Energy transition = energy mix + storage mix

… must be expanded to include the summands:

+ bridging technologies + future technologies.

The challenges of energy storage in a sustainable infrastructure in the energy transition therefore also lie in:
+ Permanent energy storage over weeks and months,
+ high storage flexibility,
+ storage as close as possible to generation,
+ development of as many energy paths as possible,
+ openness to technology.

The energy carrier hydrogen fulfills these points.

This is a very clear contradiction of full electrification in the context of the dialogs on the energy transition:

A relevant, but often neglected path in the energy transition dialog is the heat sector. <u>In fact, if we wanted to generate heat electrically, we would have to more than double the amount of electricity already generated today!</u>

However, the problem with renewably generated electrical energy is that the grids cannot always absorb the energy that can be generated by the existing wind turbines, meaning that wind turbines often have to be curtailed.
So how is the grid supposed to be able to meet the technical demands of full electrification?

At that moment, we were actually only looking at the heat, not at the aforementioned energy consumption of natural gas in Germany, for example.

At this point, one result had attracted a great deal of attention: "The cellular approach"

In June 2015, the German Association for Electrical, Electronic & Information Technologies (VDE) published a study entitled "The cellular approach" with the headline "VDE study shows how electricity grid expansion can be reduced".

Note WHO published this study!
If an association that represents the interests of electrical engineering, electronics and information technology comes to the conclusion in the study that grid expansion can be reduced, then it becomes very clear that the hurdles mentioned here are real with a view to full electrification.

The study "The cellular approach" looks at the basis for a successful energy transition across regions.
One of the VDE's guiding principles with regard to the study "The cellular approach" is how to bring calm to the heated debate surrounding the energy transition in order to use the results to drive forward the implementation of the technical possibilities identified.

The discussion between the federal government and the federal states also shows that the energy transition is wanted, but with the expansion of the infrastructure, conflicting ideas about what the infrastructure can or should look like begin to emerge.

The study "The cellular approach" is also interesting when you look at the list of authors; experts in electricity and gas technologies as well as experts in electricity and gas infrastructure were involved. The result shows the recognition that the energy transition is linking the heating, gas and electricity sectors, which were previously considered separately, with smart grids.

The development and expansion of modern energy infrastructure.

As a result, the energy infrastructure can and must be different with a view to the various challenges
+   in the urban centers
+   and in the countryside,
+   with the mobile
+   and with stationary applications.

As a result, a look at the challenges in conurbations and rural areas has shown that numerous overarching pillars of a sustainable infrastructure in the energy transition can be identified:

a)

Since the energy transition is also the path from today's material-based energy sources (energy tied up in coal, oil and natural gas) to an increasingly regenerative electricity industry, energy storage systems are also undisputedly important components in a sustainable energy infrastructure.

b)

Renewable energy generation from the sun and wind does not occur at the same time, but rather fluctuates. Both short-term and long-term storage systems complement each other in a sustainable energy infrastructure.

c)

Electrical energy is not necessarily consumed at the place where it is generated. Conventional power plants generate energy in line with electricity consumption in the grid. With modern technologies, energy generated from renewable sources can be fed into the grid and operated like a virtual power plant. Energy storage systems, such as hydrogen, can also be used to store the important base load in the grid of the energy supply companies, also with a view to security of supply.

d)

Modern technologies open up the possibility that not only the generation and supply of energy must follow the demand of consumers in the grid, but that some electrical applications can/may follow the stronger energy generation from wind and sun. Modern technologies can therefore have a price-reducing and price-stabilizing effect for energy consumers. For example, if a washing machine is pre-programmed before leaving the house and is given priority to start when more renewable energy is fed into the electricity grid and starts later, in any case at a set time at the latest, so that the laundry is actually washed when the machine returns home. Electric night storage devices can also charge first when more renewable energy is fed into the power grid and charge secondarily at the usual times.

Many cross-industry companies have already formed strategic alliances.

A central question of what the energy supply will look like in the future has also been brought to the attention of those who have not previously wanted to deal with this issue.

Highlighting the unique selling points is therefore important both for municipalities (regions) and for companies in competition with each other.

Who wants to invest in a future with uncertainty? Who wants to invest in the energy transition?

Which part of the population in Europe should want to invest in the implementation of the energy transition if the deliberately unsettling statements about the named drivers for the implementation of the energy transition are spread?

The business community cites the lack of consistent and, above all, coherent political guidelines as the trigger for their hesitant implementation of the energy transition. In fact, few people in Germany still believe this.

In view of the points mentioned in the course of the book, there is great astonishment and even incomprehension among the population as to why these promising drivers of a sustainable infrastructure in the energy transition have not yet led to its implementation.

The political developments caused by the Russian military invasion of Ukraine in February 2022 have triggered a global reorientation.
A German political statement since then is:
"Hydrogen is freedom technology."
This statement was made on February 24, 2002 by the German Chancellor Olaf Scholz (SPD) and the German Federal Minister of Finance Christian Lindner (FDP).

The drivers for implementing the energy transition are well known:

+ the expansion of renewable energy generation,
+ the simultaneous reduction of fossil fuels.
+ Energy grids are becoming flexible with the sector coupling of electricity with gas and heat,
+ the sector-coupling energy carrier hydrogen.
+ The increase in energy efficiency,
+ the simultaneous development of components and appliances with lower energy consumption.

With a view to these recognizable connections, for which the energy transition policy had not jointly enabled the necessary framework conditions, it became clear that in the course of the election campaign for the 2021 German federal elections, people no longer wanted to follow the statements of those who had been politically responsible for the energy transition until 2021.

The Federal Ministry of Economics, led by the CDU, had clearly acted against the formulated goals of the Federal Ministry for the Environment, led by the SPD.

Yes, the energy transition is not trivial, people in Germany have recognized that. All the more reason for many people to take a very close look at who they entrust with the urgent implementation of the climate targets. In view of the effects of climate change and the environmental problems, people in Germany have no justifiable reason for continuing to delay the implementation of the energy transition.

The booming global image of the energy transition

In the course of the book, it is shown that and how the energy carrier hydrogen creates the bracket that encompasses the energy transition together with climate change, environmental protection, energy supply security ... all the way to world peace.
All the way to world peace? How is that possible?

On the one hand, the energy transition has the goal of increasingly avoiding the burning of fossil fuels.
Hopefully, wars over fossil fuels will soon be a thing of the past. The global focus on wind and solar energy is not as concentrated on regionally productive parts of the world as it is on the availability of fossil fuels such as oil and gas, so that no new geopolitical power factors will emerge.

Renewable energies such as wind, solar, geothermal, hydropower, biogases and the entire spectrum of renewable energy production have the potential to make the world a more peaceful place.

On the other hand, halting the effects of climate change is also important with regard to countries in Africa, for example, where life is already difficult for the people living there due to poor soil and a lack of water.

If climate change makes life impossible for people in these regions and millions of people make their way to the more temperate zones of the world as a result, then world peace will be threatened.

China has also recognized these connections; China less with a view to a possible migration of peoples triggered by climate change, China rather with a view to Africa as a future location, which offers enormous opportunities for expansion with the potential of renewable energy generation from the sun. China is already lending a lot of money to African countries; money that the African countries will presumably not always be able to pay back in view of the contracts, meaning that the infrastructure development promoted by China, such as the ports in Africa, may become the property of China in part or even in full.

It is currently clear that China is building up geopolitical power by maximizing its know-how with a view to future technologies and by striving to gain a share of global infrastructure.

However, it is unlikely that a country with the economic potential of the energy transition can achieve such great global power that it can virtually paralyze the world with an energy embargo, as OPEC did with the oil embargo in 1973.

Should a country trigger a trade war, the renewable energy plants that have already been installed will continue to generate energy.

The EU had already emphasized in the millennium year 2000 that the time had come for a uniform European energy policy.

Africa is not the only country to be mentioned when talking about renewable energy production as an export commodity. The focus is also shifting to countries such as Iceland and Norway, which are increasingly being referred to as the Kuwait of the day after tomorrow in analogy to their wealth of crude oil.
Iceland generates a large part of its energy from geothermal energy; the geysers of Iceland demonstrate the enormous availability of energy. Iceland needs only a fraction of the geothermal energy for its own energy requirements.

Norway generates a lot of energy from the wind; the average annual wind speed in Norway is 38 km/h.

In Russia, in the USA, in Europe, in Africa and also in Australia, the potential for renewable energy generation is known all over the world.
The hydrogen energy storage system is the global link that connects the globally distributed renewable energy sources with the technological expertise and infrastructure know-how.

Furthermore, the energy transition has the potential to ensure that prosperity in the world, which is essentially linked to access to energy, can be distributed more fairly.

Energy supply and healthcare - how do they go together?

With the following link, the EU documents the increasing influence of mankind on the climate and also on the temperature on earth, triggered by the use of fossil-based energy, by the cutting down of rainforests and by livestock farming.

With this link, the EU also documents the causes of climate change and the consequences of climate change and, in addition to the dangers for the plant and animal world, also names the dangers for human health.

The EU's climate policy:
https://ec.europa.eu/clima/change/causes_de

In issue 3 in September 2013, the UMID Environment and People Information Service presented the key topic of the energy transition and health with the participation of the German Federal Office for Radiation Protection, the Federal Institute for Risk Assessment, the Robert Koch Institute and the Federal Environment Agency.

It was emphasized that the measures of the energy transition can help to reduce the burden of disease. It was also pointed out that in the past the cross-cutting issues have not been dealt with in such a way that the cross-sectoral results cannot be shown, such as the objectives of the energy transition with the recognized health aspects.

This is precisely the reason why citizens have still not fully accepted the energy transition.

As a result, the energy transition is actually a renewable energy mix combined with a storage mix, which is completed with the sector mix: electricity with gas, with heat and with mobile and stationary user coupling.
Such a complex combination without a holistic view has not been able to create acceptance in the past with regard to the skeptics of the energy transition; such a complex combination requires a high degree of practical demonstrations for citizens.

Decarbonization, the renunciation of burning fossil fuels such as coal, oil and natural gas, reduces air pollutants; this has been easily explained to citizens. The policy was able to trigger a great deal of acceptance, because burning fossil fuels is also a major cause of lung diseases.
However, there has already been a lack of communication regarding land conflicts with regard to cultivation in the agricultural sector; the production of bioenergy from renewable raw materials such as energy crops (maize, etc.), such as wood, would be in competition with the cultivation areas for the food industry and would also endanger the areas of nature reserves.

Generating the same amount of electricity from bioenergy actually requires a hundred times more space than generating this energy with solar modules.

Wind turbines do not create any relevant land conflicts, e.g. with regard to agriculture.

The image of the energy transition has improved in recent years.

While some twenty years ago, the lobby for fossil fuels such as coal, oil and natural gas argued that renewable energy could not provide a base load due to volatile, fluctuating generation, conventional power plants were presented as having no alternative.

This argument was invalidated with the hydrogen energy storage system.

About ten years ago, the lobby for fossil fuels such as coal, oil and natural gas argued that renewable energy, even with hydrogen storage, would be far from sufficient to cover the high energy demand in Germany, the grid centrality of conventional power plants was emphasized, and unjustified grid problems were cited with decentralized renewable energy generation.

Projects such as the hydrogen user center h2herten have shown that regenerative energy generation can be used to manage systems that guarantee a secure supply as an isolated solution, while at the same time having a stabilizing effect on the grid with parallel operation and intelligent grid management; the virtual power plant.

Just five years ago, the lobby representatives of the fossil fuels coal, oil and natural gas argued that the challenges of the energy transition were not yet technically mature due to the lack of a serious and holistic discussion on the energy transition.
The fossil fuel lobby has countered this by deliberately posing incomplete questions to create uncertainty.

These are positions that we encounter several times in the course of the book with regard to the subject areas and the holistic view from energy generation to applications:

We often heard it said that we first have to find out whether electromobility with batteries or electromobility with fuel cells will prevail.
It is true that vehicles with batteries and vehicles with fuel cells complement each other in electromobility, just as vehicles with gasoline engines and vehicles with diesel engines currently complement each other in combustion mobility.

It has often been said that security of supply with electrical energy can only be possible with grid centrality.

It is true that renewably generated energy with hydrogen not only enables decentralized stand-alone solutions, but also enables grid-parallel systems that can be coupled with each other through intelligent grid management to form a virtual power plant.

It has often been said that Germany is an energy importing country and that energy in the form of hydrogen will also have to be sourced from abroad in the future because Germany cannot provide enough renewable energy.

It is true that hydrogen from abroad with its transportation infrastructure is considerably more expensive, that with the expansion of renewable energy generation through modern wind turbines in conjunction with the generation of electrical energy through modern solar plants, a considerable proportion of hydrogen production can take place in Germany and can have a price-reducing effect for energy consumers.

Every kilowatt hour of renewable energy that is generated locally or regionally in Europe contributes to reducing energy costs by avoiding global energy transportation routes!

The ancillary costs for housing through to the production costs in the economy are linked to the price trends for energy.

Acceptance of the energy transition has reached all sections of the population in the EU with regard to its necessity.

How both private interests and industrial goals, as well as municipal (regional) opportunities can be shaped with regard to investments, is currently developing potential.
Where the public sector is leading the way and where the energy transition can be experienced and experienced, sustainable regional competencies are already evident today.

The creation of cooperative structures between municipalities and regions has shown that the current transition of hydrogen technologies from research and development to the market has created numerous opportunities along the value chains, starting with energy generation, energy storage, energy distribution and energy supply through to applications, whether stationary, portable and/or mobile.

Electromobility, digitalization, climate change, environmental protection, structural change ... these complementary topics must no longer be considered separately.

The success of the European economy in international competition depends crucially on the start and speed of the transformation processes in all company structures, which presupposes the willingness to cooperate and the transformation of continuing vocational training in companies.

North Rhine-Westphalia has internationally recognized expertise not only in hydrogen and fuel cell technology, wind energy, photovoltaics and biomass.
Industrialized countries around the world, increasingly also the USA, as well as China, India and Brazil, make use of our energy expertise.
Small and medium-sized enterprises have taken on an increasingly strong role in the development and production of future-oriented energy technologies with the corresponding service offerings and services. The excellent qualification of skilled workers, the human capital, is another key to the success story in NRW.

The question of why the implementation of the energy transition did not start a few years ago makes it important to look at the processes in the economy to date.
In the past, the results of basic research and materials research tended to be captured by innovative, mostly medium-sized companies.

In the past, the results of basic research and materials research were also taken up by company founders who used the results to build up a new business area.

In the past, the resulting entrepreneurial results or products were bought by the large companies; in some cases, the SMEs were taken over by the large companies, e.g. in view of the patents already obtained.
In the past, large companies were referred to as lead companies because they were able to exert a significant influence on the markets.

With a view to the energy transition and with a holistic view from the source, from renewable energy generation, to the sinks, to the applications, a whole "bouquet" of entrepreneurial opportunities has emerged.

Although the diversity of opportunities presented by the energy transition has been illustrated in terms of how it works, the diversity of opportunities presented by the energy transition has not been illustrated in the time-limited considerations of the balance sheets of large companies, whose shareholders focus primarily on maximizing the profits of existing business areas.

This has not only led to a situation where the energy transition did not fit into the target agreements of large companies, which usually have a term of one or two years, but also to a situation where, from the shareholders' point of view, the energy transition could create an economic sector that could compete with the current, existing business areas in the future.
This is due to their product-oriented approach, which has often not allowed for a holistic view.

If this was the case in the past, how did the first energy transitions from burning wood and coal to burning oil and natural gas succeed? ... From oil and natural gas combustion to nuclear energy?

Answer: Via the niche markets.

Wherever an optimization has been found in the processes, e.g. to simplify workflows or to achieve the desired results, inventions and developments have been integrated into existing processes; initially in a very special application, in a market niche.

Niche markets are not temporary!
Niche markets are the first entries into real markets.

Even the most successful developments on the market have their origins in a special segment with an initially incomplete value chain.

Expanded market models or new and complementary market models emerge at the tangents of niche markets.
In the past, regions have been able to successfully create numerous jobs by recognizing and establishing entrepreneurial activities in niche markets.

Recognizing a market niche does not necessarily result from a supply or market gap.

Recognizing a market niche can also result, for example, from the attractiveness for consumers (mainstream) or, as in the case of the energy transition, from an environmental policy necessity.

But why is it now the globally active energy supply companies that are starting the current energy transition with hydrogen energy storage?

Because a sustainable infrastructure in the energy transition links the existing electricity, gas and heating sectors with smart grids, creating a market environment that allows the existing market mechanisms to transition into bridging technologies that pave the way for future technologies.
<u>The energy carrier/storage medium hydrogen is a supporting pillar here.</u>

A close look at the technical potential of power to heat, power to fuels, power to gas and power to chemicals with a view to coupling the electricity, gas and heat sectors in smart grids reveals material flows at whose tangents further potential is emerging.

Hydrogen energy storage is playing an increasingly important role in more and more technical potentials; when produced from renewable energy, it is also environmentally friendly and climate-neutral. In order to recognize and understand these interrelationships, it is interesting to look back into the past, as this reveals the trigger for the past lack of dynamism.

Is it really so easy to point the finger at large companies in the energy industry or at vehicle manufacturers?
No.

The goal of a company should be to maximize profits. It is perfectly legitimate for a company to protect this goal in competition with other companies and other products. Competition is not a threat in the markets; competition is often the trigger for innovation.

Why the current energy transition is only now being communicated as an innovation and no longer as a threat to the markets is not solely due to environmental policy necessity.

The reason for the past lack of dynamism in the energy transition is not just that a whole "bouquet of flowers" of entrepreneurial opportunities with regard to the energy transition and with a holistic view from renewable energy generation to applications would have been confusing.

The reason for the past lack of dynamism in the energy transition lies in the combination of the diverse opportunities identified by a diverse corporate landscape, which were focused on their own business areas that had previously been considered separately.

The famous "thinking outside the box" approach quickly touched on other business areas, meaning that further communication did not take place.

This list is provided for clarification:

The key technology for the success of the energy transition is, on the one hand, hydrogen technology with the storage and provision of energy.
On the other hand, the key technology for the success of the energy transition is energy conversion, which also includes fuel cell and electrolysis technologies.
So far, this is technically understandable.

In the past, the orientation of the companies was different with regard to their entrepreneurial responsibilities.

The key companies for the success of the energy transition are numerous:
+   the energy suppliers,
+   the chemical industry
+   plant construction,
+   mechanical engineering,
+   the automotive industry,
+   and many more.

One would think that if an entrepreneurial opportunity arises with the energy transition through the energy storage system hydrogen and through the energy converters fuel cell and electrolysis, that an overall interest of the companies would "naturally" arise from this; that the chemical industry would therefore enter into a dialog with plant engineering with a view to hydrogen, with a view to fuel cells and electrolysis.

Thus, with a view to the complex interrelationships mentioned here, it is also recognizable in the energy transition that once the opportunities have been recognized, one simply takes action and starts one's own initiative.

The regional nucleus in the Emscher-Lippe energy region is h2herten, which has quickly gained supra-regional importance and is still one of the rare centers in the EU that successfully demonstrates THAT and HOW the energy transition works.

The hydrogen user center h2herten:

On the one hand, the international flow of visitors and, on the other, the international cooperation with companies from Japan, Canada and the USA confirm the success.

Resource efficiency and market trends justifiably show that the desire for an energy transition can also become a reality.

With a view to energy consumption, resource efficiency is above all also the endeavor to reduce energy consumption, e.g. with the same work performance of machines, both mobile and stationary.

Resource efficiency can be supported by using energy more consciously and sparingly, which includes heat as well as electricity. However, resource efficiency is also the endeavor to use considerably fewer raw materials. The recovery of raw materials will shape sustainability in the future.

Reliability, convenience, prosperity, sometimes even just the mainstream, are triggers for market trends.
What is currently influencing market trends is the recognition of the impact of climate change.
The recovery of raw materials, recycling, is becoming very important with the success of the energy transition:
+	That cheap products have to be purchased more frequently as a result due to a lack of material quality and are more expensive as a result than the purchase of high-quality products,
+	The fact that cheap products result in an increased amount of waste has reached the public's awareness.
+	Where low-cost appliances are preferred, there can be no question of energy-saving use; the components used in low-cost appliances are not designed for energy efficiency.

Prosperity and convenience are therefore not at odds with the desired reduction in the consumption of raw materials; on the contrary, as the past shows us, innovation has very often led to greater resource efficiency and greater convenience at the same time.

Compared to classic cars, modern cars show that the desired resource efficiency and comfort go very well together.

The job engine of the energy transition

Suddenly, skilled workers are increasingly in demand again; companies are increasingly recognizing their employees as imaginative and innovative again: the return of human capital in the perception of rethinking entrepreneurs.

The skilled workers needed for production companies are available on the labor market in Germany due to the sharp decline in heavy industry.
The current challenge facing the economy is the provision of suitable training places and the appropriate retraining and further training of employees.

I would be delighted if we could come to a common understanding that the energy transition, with all its exciting and at the same time promising sectors, is the decision for sustainable cooperation.

The author

Dieter Mende

Homepage:
www.eez-mende.de

The book is written with the background of professional training in chemistry as well as in electrical engineering, supplemented with the background energy dialog EEZ energy energy industry future energies; I myself am the founder of the energy dialog EEZ on 05.07.1995.

My professional background is wide-ranging:

Commissioned since September 1997 with the central control technology for energy centers in the automation technology project office of a regional energy supply company; in November 2019, I moved to the electrical power grid construction project planning department.

Since February 2003, initially commissioned with the tasks of setting up and expanding the regional hydrogen nucleus h2herten, then commissioned with tasks of project and company acquisition, market communication and net-working in the team of the hydrogen user center h2herten.

In both cases, the employers noticed the very successful EEZ energy dialog, which led to the jobs.

My motivation for creating reports and books is, on the one hand, a passion for highlighting the opportunities and possibilities in the potential grid of the energy transition with hydrogen as an energy carrier and, on the other hand, the ambition to establish and expand a hydrogen infrastructure with the advertising of cross-industry service providers, with the identification of sustainable contributions and the resulting complementary competencies.

I met the established companies in the energy market and their initial rejection of hydrogen as an energy source and the fuel cell as an energy converter with meaningful results of the potential analyses, with the potential of the Power-to-X sector coupling, the identification of unique selling points and with the complex project impulses of the multi-layered interests of all those involved.
With perseverance and patience, even initial skeptics of hydrogen as an energy source and the fuel cell as an energy converter were successfully won over to a sustainable infrastructure in the energy transition.

With the EEZ Energy Dialogue, I am a long-standing member of the DWV German Hydrogen and Fuel Cell Association. The DWV is one of the successful mouthpieces in Europe for hydrogen as an energy source and for the fuel cell as an energy converter; the DWV speaks with the results of over one hundred industrial and research institutions.

I successfully contributed to the acquisition of members for the foundation of an advisory board for h2herten, from which the advisory board for the h2-netzwerk-ruhr emerged through expansion in 2008.

Picture above:
The hydrogen user center h2herten: the regional hydrogen nucleus in the northern Ruhr area.

Picture left:
14.06.2019: The opening of the hydrogen filling station at the h2herten user center;
in the background the former coal mine
Auf Ewald in the background.

Pictures: Dieter Mende; EEZ
Energie Energiewirtschaft Zukunftsenergien